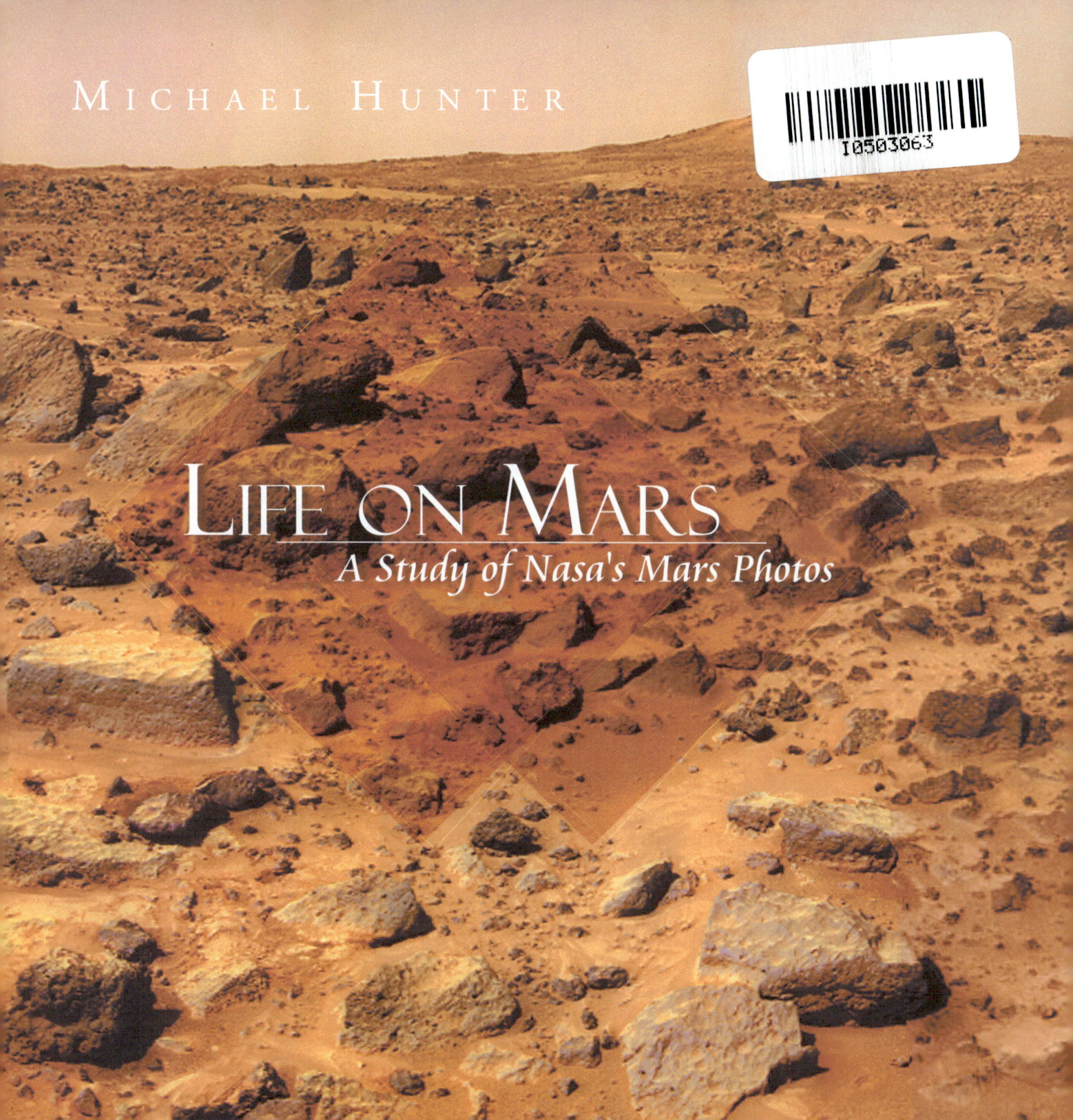

MICHAEL HUNTER

LIFE ON MARS
A Study of Nasa's Mars Photos

LIFE ON MARS

Michael Hunter

CONTENTS

INTRODUCTION

As I work on writing this book, NASA's Curiosity Rover is examining the terrain in Gale Crater on Mars. It's objective is to find evidence of microbiologic life. The rover scoops soil, touches rocks, does spectrum analysis of rocks by shooting laser beams at them, and it takes pictures which it transmits back to Earth. Last week, it discovered an ancient streambed, with rocks that were worn smooth by rushing water. However, in two months of exploration, at this time, NASA has not mentioned any water, plants, or life forms that I am going to show in this book. For scientists, today, Mars is lifeless and uninhabited, with temperatures too cold, and atmosphere too thin to support life.

For ages, humans have wondered if there is life on other planets. NASA's Mars rover photos show that indeed, Mars has many life forms, some of which are identical to those on Earth, but in addition, many which are completely unknown by Earthlings.

In early July, 2012, I just happened to see an article on the internet titled "New NASA Mars Images". As I looked over the images, I began to wonder if there are life forms on Mars, and if so, if maybe there are some kind of traces of them, such as tracks in the sand or snow. I looked and looked, and although I saw something like tracks in the snow, I had to conclude that it wasn't necessarily animal tracks and it would be impossible to prove even if it were. It might be something geological.

When I was quite convinced that there was no visible evidence of life to see, near the point of abandonment of the task, my eyes fell upon a picture of a circular, white disc in the orange Mars

sand. It looked just like a sand dollar. That was the event, the moment of realization, which caused me to become fascinated with life on Mars. Subsequently, I did see, and realized, that there are birds, reptiles, mammals, dinosaurs, extinct on Earth, and still other strange, unidentified creatures, "swimming" in and walking across the soil, and flying in the air, all over Mars. What people have been led to believe to be rocks on a lifeless planet are, in many cases, heads of live animals, which apparently live under the soil due to the extreme environment, in a way similar to the way marine life lives under the seas on Earth. With this book, I'm going to attempt to share with you, and to convince you to believe, what I have learned, considered, and postulated as a result of my study.

Ask most people today if they believe there is life on Mars and they will probably respond with doubt but with some hope that there could be microbiologic life of some sort. This is the mainstream belief that prevails today. In fact, most expert scientists (astrobiologists) would say the same thing, that there is little likelihood of life other than microbiologic life, some form of invisible, living cells that would be very difficult for anyone to find without complex scientific study, such as spectrum or chemical analysis.

This book is written for those who are willing to step outside that box. This book is for those who are willing to look at the question, "Is there life on Mars?" with an open mind, clear of scientific dogma that has prevailed for the past fifty years or so. It is for people who wonder, for example, if life can adapt to minimal levels of oxygen...if life can adapt to minimal temperatures... if life can subsist in climates that are harsher than the harshest on Earth...if the data about the climate on Mars has been wrong... if the temperatures are more stable and higher under the ground... if there truly is water and plant life that is not yet clearly evident on Mars... if there are conditions that are adequate for life on Mars... if there IS truly life on Mars?

The pictures in this publication are taken from images from NASA's Mars rovers. Some people have

difficulties in interpreting the images because of the poor visibility. There are several reasons for that. One reason is that the photographic processes used to obtain the photos in the first place are complex and involved. The rover cameras are protected with dust cover lenses of plastic, which distort images to some degree. Furthermore, still photographs are difficult to interpret because obviously, there is no motion to observe, which is one key factor in seeing living objects. And also, some of the photos appear to be black and white, due to the low level of light on Mars. The images are described by NASA as mosaics, taken by multiple cameras over long periods, and then pieced together. And also, the images are sent from Mars to Earth which involves a transmitting and receiving process. All of the processes involved lead to problems in visualization, especially with photos taken at great distances.

One consequence of the mosaic process is that there seem to be multiple versions of some of the photos, which show various focuses, various lighting, various colors, and other varying factors. As an end result, interpreting the photos are to some degree, confusing. And on top of all of that, there is loss of quality due to publication and handling. Therefore, I would ask readers to be extremely flexible when viewing them. The images can be very revealing and conclusive to those with open minds who study them very carefully. Admittedly, the pictures in this book can strain the imagination and at the same time can be difficult to interpret and comprehend.

I'm writing this book because I believe the world is wrong about their beliefs regarding life on Mars, and it is clear from the pictures sent to Earth from NASA's rovers that commonly believed assumptions regarding the uninhabitability of Mars are flawed. And, because of that, there seems to be a problem with the beliefs of the scientific community and possibly, their way of thinking. Isn't it illogical and erroneous to believe that life on Mars is essentially impossible, given that NASA has taken pictures showing animals or all sorts, in several locations on the planet? One picture after another, over the past decade and more, has shown evidence of animal life. For those who are willing to believe their eyes, in spite of scientific dogma, I present this book for amusement, information, and education.

PART 1: WATER ON MARS

Is there water on Mars? Scientists call it a "dry" planet. The general belief is that there is frozen water at the poles and below the ground, but that liquid water cannot exist for any long duration. The scientists lately talk about frozen carbon dioxide on the poles, and snowflakes of carbon dioxide, traces of liquid water being just recently being discovered. But if you look at the rover photographs, and you will see that water is all over the place on Mars. In Gale Crater, it can be seen as a pool in the satellite photo at right:

In the enlargement at right, the pool at the base can be seen to reflect the area above the pool. This is convincing visual evidence that this is in fact a pool of water, rather than just a dark sedimentary geologic layer. Note waterline/ reflection location with arrows.

Curiosity has photographed views that appear to include waterways. The dark band in the photo below appears to be water. In fact, something can be seen swimming in the water. The dark muddy banks show that tidal action is possible, and the water level varies with time. NASA indicates the distance from the camera (2 km).

 In this enlargement of the lower left corner of the photo above, animals can be seen swimming in the waterway. In fact a close inspection reveals the plates on the spine of two stegosauri, and a spiked tail of another. So, it looks like water, it pools like water, it reflects like water, animals swim in it like water...perhaps it IS water.

In the photo at right, by enlarging, one can see a pipe spewing water into the mouth of an animal. For more on pipes, see PART 5.

Water is spewing from a t-shaped pipe here: (location indicated with arrow a):

An animal is lying in the pool and drinking from the water, here:
(location indicated with arrow b):

So the question seems to be not, "does water exist on Mars?", but rather, "How can it exist?" There are other questions that arise when looking at the Curiosity rover views of water. Where does it come from? Why does it not evaporate? I cannot explain those things, but I can guess. My guess is that the water came from an asteroid. Gale Crater must have been caused by an asteroid which contained water...a giant snowball. The ice must have slowly melted into the ground and it essentially freezes daily, or more precisely, nightly, which slows its evaporation. Imagine the freezer in a refrigerator (non-frost-free), which accumulates ice from condensation which needs to be thawed periodically. Upon thawing, the ice becomes water which then vaporizes. As time goes by, the water in the air condenses and refreezes. So, even though the water is constantly changing state, liquid, ice, vapor, it is locked into the crater vicinity. I believe it evaporates into the meager atmosphere, and then condenses during the nighttime low temperatures, on the surface. A similar process may keep the "deserts" moist under their surface. It's just like a freezer in which water vapor condenses on the solid matter surfaces.

PART 2: PLANTS ON MARS

Animals cannot survive without plants, and plants cannot survive without water. Those are facts of life on Earth. This may seem simple and common knowledge, but this bit of information is quite important in the question of whether there is life on Mars. Where are the plants on Mars? Obviously, on Mars, plants are difficult to see. If they are there, then they are not showing up visibly, at least significantly, on photographs. The lower light levels, higher carbon dioxide levels, and lower temperatures on Mars suggest that plants, if they exist, are able to withstand those conditions. Certainly, plants on Mars could and should be different from those on Earth, or possibly, they are similar to some of the desert plants on Earth.

Animals need plants for a food source. if there are animals, there have to be plants. So maybe on Mars plants grow under the soil but near the surface? Are plants on Mars subterranean? A plant in water gets light through the transparent water, while a rock under sand cannot, because the rocks are opaque. Another possibility is that the plants look like rocks, or simply are difficult to see because of their color and shape, and the limitations of the rover photography.

In NASA's rover photos there are plants which look a lot like rocks in the rover photos. Consider the photo at right, from the Phoenix rover in May, 2008. Notice that several flowers and plants are visible. Are they petunias and lilies? Wandering Jew? A mushroom?And is the flat, red leaf ground cover a type of Coleus?

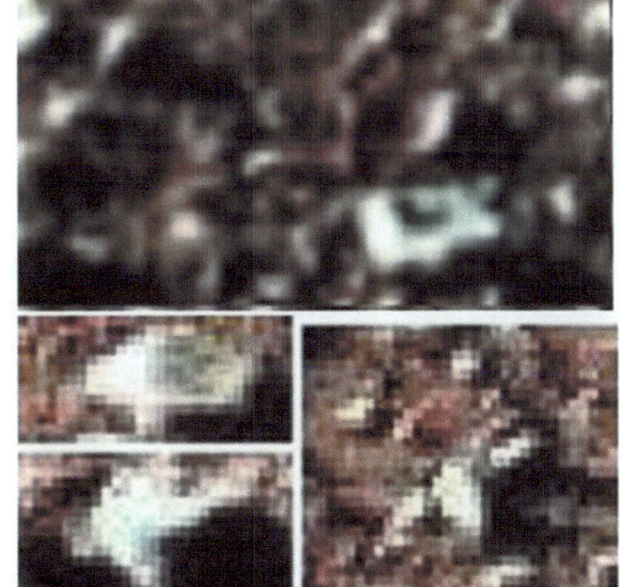

The right photo was one of the first that was sent by Curiosity rover when it landed August 6, 2012.

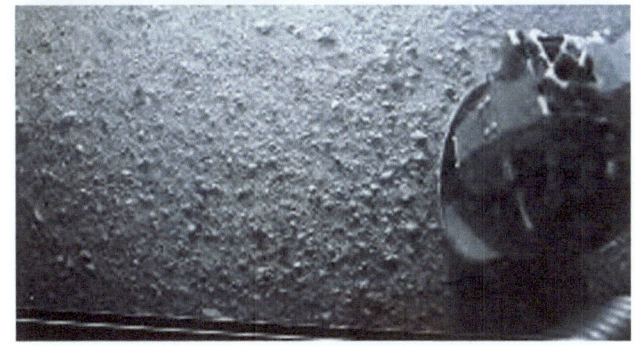

15

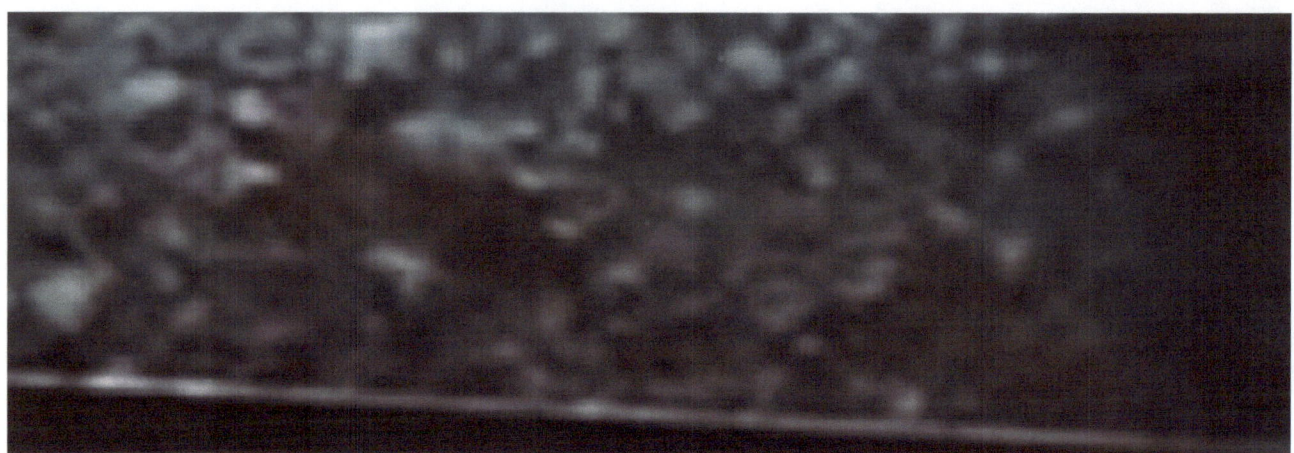

Looking at the lower right portion of the photo, below the rover's wheel, you can see a small red area in what otherwise looks like a black and white photo. Those are red leaves of plants, possibly coleus.

Looking at a closer view of a portion of the overview photo, you can see what look like pansies. The lighting is very dim, causing them to appear to be white.

This demonstrates that plants DO in fact grow on Mars, in spite of the severe temperatures and thin atmosphere. The problem in seeing them, it seems, has been visibility. The photos from Mars simply need better clarity and enlargement to see plants more clearly.

PART 3: EARTHLIKE ANIMALS ON MARS

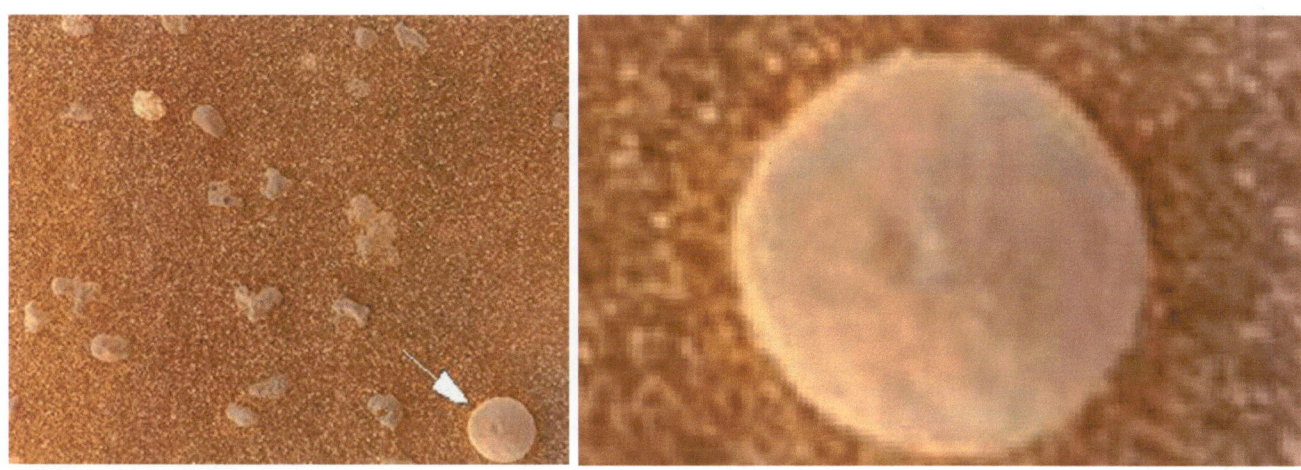

SAND DOLLAR

It was actually quite astonishing for me to discover a sand dollar on Mars! The photo from the Mars Exploration Rover Opportunity, released on February 4, 2004, had been available for eight years and nobody ever mentioned it publicly, to my knowledge, as showing evidence of life. Questions came to my mind. Why are there no holes in the sand dollar, like those in the skeletons of sand dollars on the beach? Why is there no clear five-petal pattern like those on Earth? Could it be that it is not really a sand dollar? My research led me to the following considerations.For one thing, I cannot dream up any alternative explanation. Could it be a circular rock worn smooth by the wind

or water? No. Show me another and I might consider it possible. Could it be a circular rock cut out by the rover and dropped on the sand? No, the rover brushed circles in rocks, but did not take core samples, and even if they did, they weren't white like that. If it looks like a sand dollar, same color, same shape, and same size, then one ultimately, with no alternative explanation, needs to come to a realization that it IS a sand dollar. Sand dollars, on Earth, when alive, are covered with cilia, tiny hair like threads that help to filter out micronutrients from the seawater and to move through the sand and water. The reason there are no holes on the sand dollars on Mars, and the reason the five-petal pattern cannot be seen, is that the cilia are still in place, (and the photo is blurry). The sand dollar may be dead, as you can see it has a fracture near the center, (which seems to occur with some regularity as well on Earth). But the fact is that the cilia are still in place, which indicates it died recently before the photograph perhaps weeks, months, or years, as opposed to millennia or millions of years. This sand dollar did not die a million years ago and retain its cilia intact for all that time, in spite of exposure to wind and weather and sun...it died recently, and NASA photographed it in 2008 and there it is for us to see and recognize and observe and enjoy as evidence of life on Mars. For a sand dollar, as lowly as it may seem in the ladder of life, must consume other living organisms to survive. And therefore, the search for microbiologic life on Mars, which NASA is about to start, is already completed, with positive results, by simple acceptance of reality of a sand dollar, and logic. There is some question as to the timing of evolution of the sand dollar. The modern sand dollar, (and this Mars sand dollar seems to match modern ones), evolved in the Eocene period, 56 to 33 million years ago. That implies an Earth-Mars exchange during that period or afterwards.

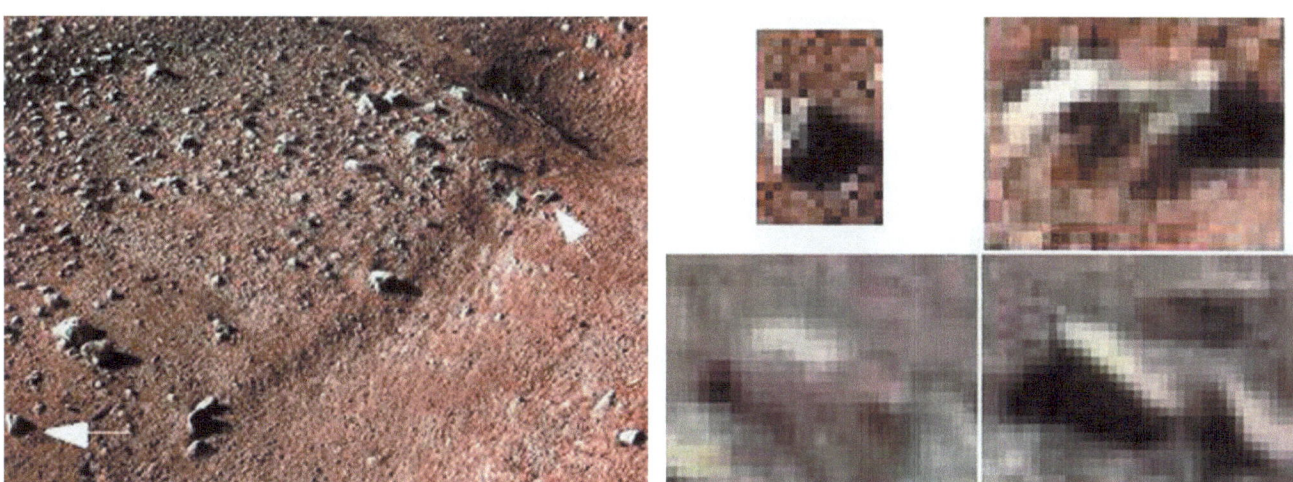

DUCKS/GEESE

I proceeded to look at another NASA photograph, from the Phoenix Mars Lander on May 25, 2008. Knowing that life on Mars was possible, helped by the sand dollar discovery, adding a bit of positive reinforcement to my study, I soon found a duck head, which looked pale and dead, but fully intact. This was unquestionable, not just something highly probable or very possible. I had no doubt. A rock cannot form just like a duck head with eyes and bill and shape of head. And to confirm my finding, in the same photograph, there was another just like it. But both of these things, the sand dollar and the duck heads, did not cry out that they were alive. They appeared dead, bleached by sunlight, perhaps recently alive. A still photograph can't be expected to "quack" or to show movement or life. And the mystery presented itself, about how could two ducks die and have their heads above the surface of the soil, at the same time, as if looking at the rover as it rolled by? After all, if it looks like a duck, same shape, same size, one needs to come to the realization, eventually, that it IS a duck.

What hit me soon after this discovery is that now, within just a few hours of effort; I had identified two species of animal on Mars that were identical to species on earth! I had just begun.

Another question that came to mind was this: "What is it about ducks and sand dollars that might give them survival ability on Mars?" Answer (possibly): both of them feed underwater. Ducks on Earth can often be seen feeding off of plant matter under the surface of the water. Maybe they do that as well on Mars. But why would ducks live under the soil on Mars? Answer (perhaps): because there is water, (and plants), under the soil! Considering that gravity is weaker on Mars, compared to Earth, perhaps the soil is not so difficult to penetrate, on Mars. Perhaps being under the soil on Mars is like living in sand underwater on Earth. It must be possible to "swim" and live, in the sand on Mars, something like clams do on Earth. Ducks, sand dollars...I soon began to wonder, what else might I find on the rover photos?

RAT SNAKES

There was something on the same photo from the Phoenix Mars Lander on May 25, 2008 that looked like an animal head poking up out of a hole. In fact there were two or more. I didn't know what this was until Curiosity sent back its first photo with a similar shaped head, like a rat with ears. I think both of those are rat snakes. The rat snake has something like ears, behind its eyes, and the head resembles a rat. Also, the jaw is somewhat flat.

FISH

In the same 2008 photo, there is a relatively clear picture of a goldfish. Close examination shows several, but some are blurry and hard to distinguish, due to their squirming, I 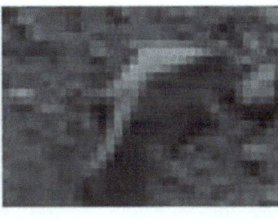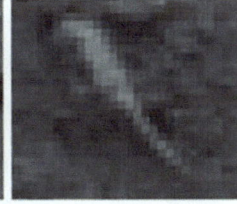 suppose. Curiosity photos show that they lie on the ground near the tracks of the rover. So goldfish live out of water while animals live underground? That hardly makes sense. There must be water, (and nutrients), under the soil.

WEASEL

This weasel, popping it's head and neck above the surface, is apparently struggling with what looks like a daschound. Both of these, incidentally, are burrowing animals by nature.

FOX

The January 19, 2005, NASA's Mars Exploration Rover "Opportunity" photo of the meteorite on Mars has this fox pup sitting in one of the cavities of the meteorite. Actually, there is some kind of animal in almost every pocket. However, it is hard to see what they were. And there are some animals in the distance, too blurry to make out their species. Foxes, again, could burrow in the sand to protect from the night cold and predators.

BIGHORN SHEEP

Proceeding to look at another photo, I focused my attention on a photo by Exploration Rover Spirit, released by NASA's Jet Propulsion Laboratory on January 10, 2004. Scouring this photo taken in Gusev Crater, I first noticed on the left hand side, the head of a bighorn sheep, which seems to have its eyes focused on the rover as the picture was taken. Now, this is something spectacular. A sheep is a mammal, further yet up the animal ladder, much more developed than the lowly sand dollar and the duck. And it appears to be swimming in the sand. It's not a skeleton or skull, but a live animal, for certain! This shows something important. Animals live on Mars today that are able to breathe the atmosphere and survive the cold nights, and to consume some kind of food, and to reproduce... animals similar to mankind! Somehow the inhabitability of Mars must be brought into question, for here, right before our eyes, is a living mammal, swimming, breathing, reproducing, and alive, in the red sands of Mars. My guess is that this sheep, and most of the animals on Mars, live under the soil during the cold nights, and they emerge when it's warm enough.

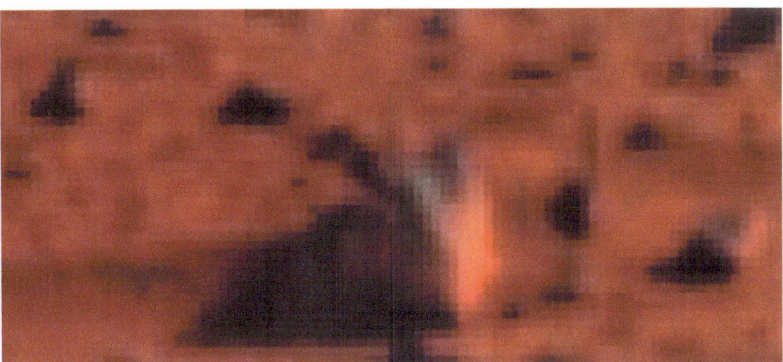

GIANT PANDA

On the same 2004 photo, just a few feet to the right of the longhorn sheep, is the head of another animal. At first this head looked to me like a human skull. As shocking as that may sound, it did look like one. But close comparison showed me it had too long a snout or nose to be human. Also, the black eyes were accompanied by a black nose and ears. I soon realized it is a Giant Panda. These creatures live on Earth

by consuming bamboo. I sometimes wondered what they would eat if they were deprived of bamboo. Well, something is there on Mars that is able to sustain them. The plant source on Mars is there, but not yet discovered. it has to be there. Otherwise, how could we see a head of a Panda swimming through the red sands of Mars? Interestingly, the four animals found so far are all plant-eaters: sand dollar, duck, longhorn sheep, and panda. This means there is "something" on Mars to feed them... plants. And therefore, also one has to conclude, there is water, somewhere, plentifully available but not readily visible. Perhaps it is under the surface of the sand. Or possibly, it is on the surface and it simply is difficult to see in NASA's photos.

HIPPOPOTAMUS

These two enlargements from the same photo show what look like hippos...from the rear.

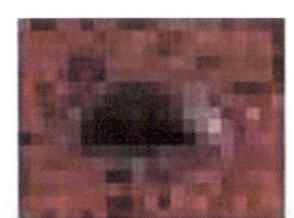

STEGOSAURUS

In that same photo in Gustav Crater, my search led to a small animal, perhaps only two feet long, mostly above the sand, in the front central portion of the photo. It has a reptile appearance, with plates protruding from its spine. The long head resembles a

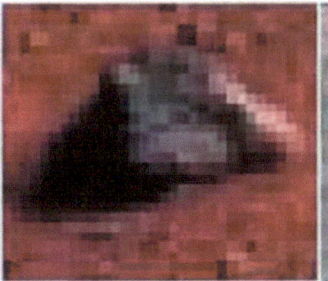
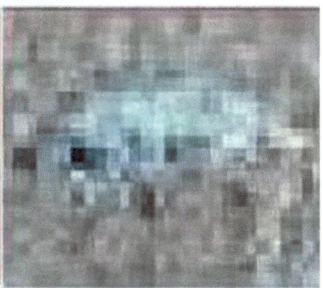

stegosaurus. There is what looks like a bigger Stegosaurus nearby, which turned its head to the rover at the moment of the picture. What looks like several "rear ends" of Stegosauruses are in many places in the photo. They look something like a house, or rocks that have peaks, like an inverted "V". It seems they all face the same direction, similar to the way cows on Earth face the morning sun.

The significance of dinosaurs on Mars that are identical to those that once existed on Earth, is supremely important. It suggests two possibilities. Either the source of life on Mars was Earth itself, or there is some external source of living animals that populates everywhere in the Universe. Either Mars or Earth at one time scraped each other, or some kind of vehicle like a comet transported life to Earth and Mars in the same manner. Which of these theories is correct, I will not attempt to resolve. However, my guess is that at one time the two planets scraped against each other and a life transfer occurred from Earth to Mars. That would probably have occurred around more than 65 million years ago, (Cretaceous period). There is a possibility that it happened more than once, considering the sand dollar. Maybe there is a 40-million-year recurrence pattern.

THE TEXTURE CAM PHOTO

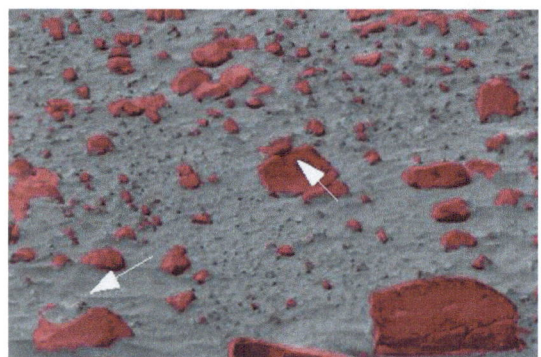

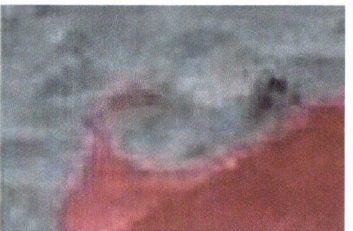

BIRDS

I don't know exactly when or where the photo was taken, but it was essentially a scene of rocks and sand, in which the rocks of a certain size were red, and the sand was grey. Supposedly, the camera was programmed to make large things red so that is could differentiate big and little rocks. But

unfortunately, it could not differentiate between animals and rocks, so in the case of a bird sitting on a rock, the bird was the same color as the rock, red. A close inspection of the birds on rocks showed one of them like a mourning dove, with a chick on its back, and the second, a pair of bird cuddled together side-by-side, heads something like vultures, with a youngling under one wing.

SNAKES

Other than Rat Snakes, there are other snakes visible in some of the photos. The meteorite photo shows something like a snake and it appears that a goose has the snake pinched in its beak. Also the Curiosity photo of Gale Crater shows many snakes, one looking like a large tan-colored boa striking a super croc. Reptiles seem to thrive on Mars.

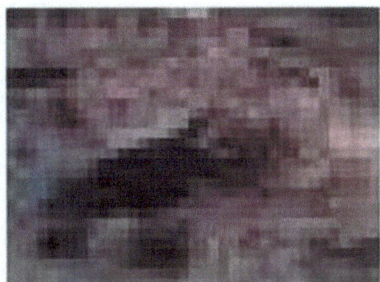

Giant Boa attacking a Supercroc Gale Crater

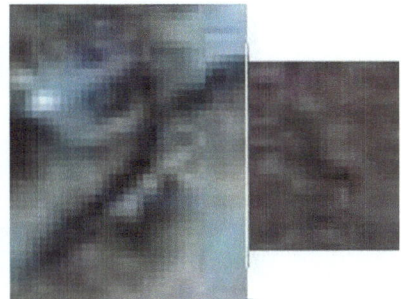

Snakes on Mars (from meteorite photo)

VELOCIRAPTOR

The triangular heads in the above photo appear to be Velociraptors which were predatory reptilian dinosaurs that lived on Earth 75 to 71 million years ago. Like the Stegosaurus, there seem to be many reptilian dinosaurs that are now extinct on Earth. And strangely, those dinosaurs lived during diferent time periods on Earth. Yet, here we see a Stegosaurus and a Velociraptor next to each other in the same photo from Mars. Stegosaurus lived on Earth 150 million years ago, 75 million years prior to the Velociraptors, according to scientists. Some of the things in these photos may cause scientists to scratch their heads in confusion.

PART 4: NON-EARTHLIKE ANIMALS ON MARS

Part 3 explained and showed how numerous animals on Mars are identical to animals on Earth. This Part 4 will explore animals on Mars that do not seem to exist on Earth. These animals seem to fall into the following broad categories: 1) extra-large, gigantic versions of known Earthlike animals, and 2) animals that seem to have never existed on Earth, and hence, are truly Mars animals.

GIGANTIC ANIMALS

Over time, my suspicion developed that some animals seem to have grown to enormous size on Mars. This awareness came to me gradually, as I noticed more and more geologic features that looked like animals. These could be giant animals that have grown hundreds of times as large as normal, or they could be species that have evolved separately from their Earthlike species. They include super geese, super crocs, and super raccoons, but of course, the effect could extend to other animals. Their giganticism may be a consequence of the planet's lack of protection from cosmic radiation from the Sun. Or, there could be some other factor which causes it. Eventually that is a question to be answered by scientific study. But it seems as if the animals grow to become super huge. Perhaps they grow bigger because they simply live longer. It seems like they eventually become as big as mountains, and they are too large to move about under their own power, therefore they simply lie around and smaller animals mistake their mouths for caves, and enter in an attempt to avoid the nighttime cold and predators. So their supply of food is huge and the energy needed to hunt and feed is negligible. That may explain how they can exist and survive in such large form.

- Flying Geese
- Supergoose
- Flying Supergoose
- Supercroc

Same photo with all but geese opaqued

SUPER GEESE

What can be said about the geese on Mars? They fly, in spite of the sparse atmosphere, as can be seen in Curiosity photos. They live under the soil, as seen in Opportunity rover photos. They also somehow grow to enormous size! The large ones seem to like to perch on mountains. They blend into the mountains almost invisibly. Even the super geese in the Curiosity photo seem difficult to see, as if camouflaged. Is the super croc trying to catch it as it flies by and is that why its mouth is open so wide? Or, possibly, the geese landed on the croc's upper jaw and were lifted as the croc's jaw lifted. The huge birdlike creatures on the hillsides and in the waterways resemble flounders, flat fish, but, they have heads like geese. My guess is that they are simply very old and grown up versions of the normal Earthlike-geese on Mars. Some snow geese seem to have buried themselves in the soil, and their gigantic heads are visible on satellite photos, poking out of the soil as in the satellite photo below. These heads are roughly ten feet across, bigger than the Curiosity rover, inset beside it for scale comparison. However, they maintain their white heads and yellow beaks like normal Earthlike Snow Geese. Normal sized geese on Mars bury themselves in soil, also, as described in Part 1.

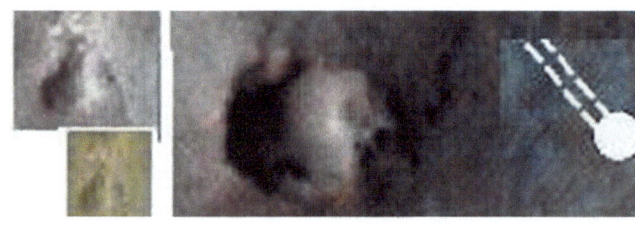

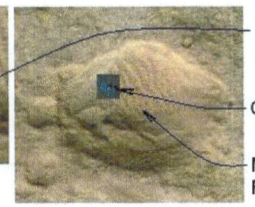

Baby
Super Racoon

Curiosity Rover

Mama Super
Racoon

SUPER RACCOONS

This satellite photo shows what looks like a baby raccoon under its mother's neck. It's like a family, cuddled together for warmth, probably. The pup is about 100 feet long, (the eye is about the size of the rover as inset from the same photo for comparison)...the parent is perhaps two times as long as the pup, judging by the size of the heads. I believe there may be two parents bundled together with the youngster, (it's hard to see precisely in this monocolored long-distance photo). This kind of gigantism seems to be fairly common on Mars. 300-foot geese, 200-foot raccoons, mountain-sized animals! What is it about Mars that causes this type of giganticism? My guess is that the lack of magnetic field on Mars allows more cosmic radiation which causes elimination of growth limits such as on Earth, but this idea begs for scientific research.

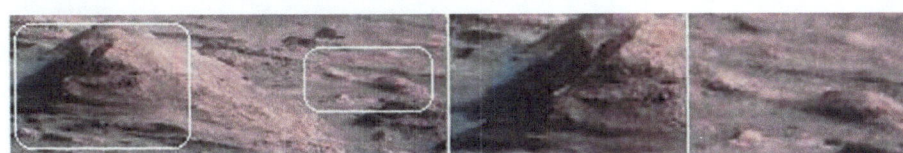

SUPER CROCS AND GIANT SUPER CROCS

NASA claims that this "pointy mound" is 300 meters wide and 100 meters tall. Yet it can be seen to have the shape of a head, including a mouth, eye, and ears and mouth. At its base are several other supercroc heads, whose bodies are apparently buried underground. The headlike "hill" does not

resemble a super croc head, precisely, because it does not seem to have a long-enough snout. It might be a super goose, but the rear part of the head is not round enough, and the bill is too short. So this might be some kind of giant reptile.

THE GIANT DOG

 The satellite photo of Gale Crater shows what seems like an animal in the center of the crater. After long perusal, I finally settled on this animal being a giant dog. I decided this because of its fur, the shape of its head, and a possible tail. This animal is (or was) huge. Gale Crater is the size of the states of Connecticut and Rhode Island combined! This shows that giganticism has very strange limits on Mars. A hundred mile long dog! Who on Earth would believe that?

NON-EARTHLIKE ANIMALS

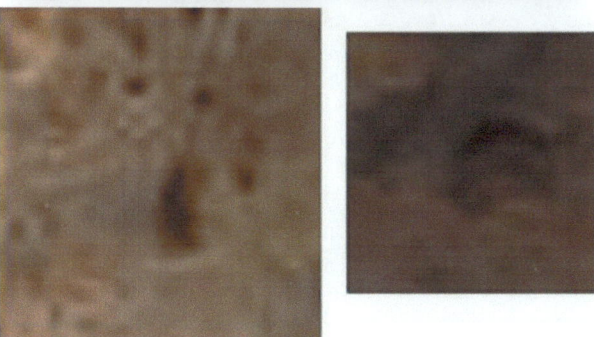

ANIMAL-LIKE BUGS ON ROCKS

This close up view of a rock (named by NASA as "Jake") by Curiosity rover shows two small things that look like bugs. They appear, by the scale, to be about two to three millimeters long, long, with elongated smooth and flexible animal-like body, multiple legs, and antennae. They resemble tiny hippos. Minihippopotamuses?

UNKNOWN SPECIES

I simply do not know what this is. Some kind of screw like dog?

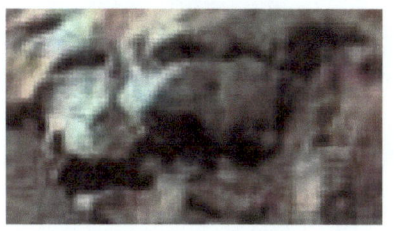

This is another animal that I simply cannot identify

2 Lizard-like animals probably 20 ft long
Curiosity photo

Similar Lizard-like animal about 2 ft long
Curiosity photo

GIANT LIZARDS

I cannot relate these large lizards to any species on Earth, even dinosaurs. The pair in the top photo appear to have just left the water and are staring in the direction of the camera, from a great distance away...perhaps a kilometer? The one on the bottom, taken by Curiosity on its first day,

was relatively close to the rover, perhaps a meter or two distant. So the ones on the top are huge, like ten meters long, while the one on the bottom is only about one meter in length. They appear similar, but not identical.

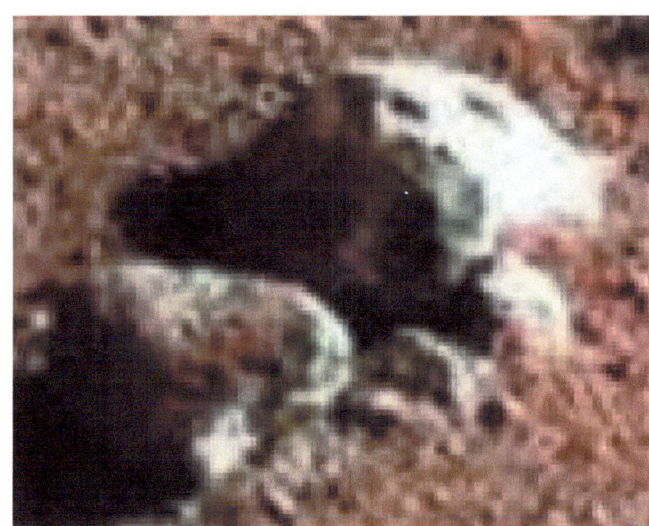

MANTIS PREDECESSOR?

The head of this unknown species is much like that of a praying mantis. However, it does not have the long slender neck or body. Mantis evolved from cockroaches. Perhaps this is in-between? It appears to be about a foot long.

34

JAKE

Is "Jake the Rock" Really A Rock? The pyramid-shaped "rock" that NASA named "Jake" and experimented with looks strange for a rock. It has a shape similar to the "church". Rocks are hardly uniform in shape, and rarely pyramid-shaped. The photo of Curiosity touching Jake shows what looks like eyes on the face opposing the rover. Also, in the photo captured by the Spirit rover in 2004, there is a very similar-looking creature in the foreground, on the left side of the photo. The body looks similar to Jake's, and it's head resembles the creature in the movie "ET". A few similar shaped "rocks" are in that photo, and so it seems likely to me that this is a species of animal on Mars, which is something like a turtle, with a shell shaped like a pyramidal rock.

  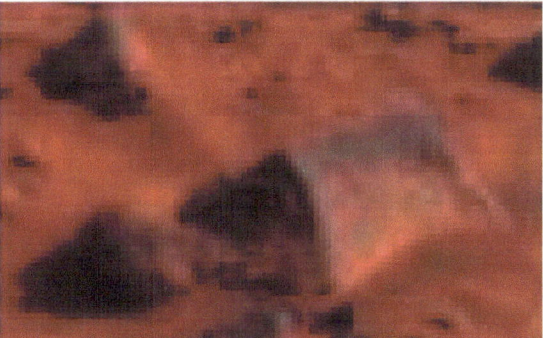

PART 5: HUMANS ON MARS

ARE THERE HUMANS ON MARS?

Why am I asking this question? Because, I am seeing evidence of humans on Mars and I need to share it with others.

If looking for evidence of "humans", we need to define, first, what precisely are they? Human beings are evolved primates that have a developed brain which allows complex thought. Beyond that, when looking for "humans" on Mars, one has to allow some flexibility in the definition. We could say that we are looking for intelligent beings, and leave the physical form "tbd". They could be similar to Earth humans, or they could be less evolved, like early man, or they could be more evolved, like the "greys", or extraterrestrials that have been witnessed on Earth. Considering the giganticism of other animals on Mars, we might find gigantic humans. Or, considering the tiny hippos that live on rocks, we might find tiny humans. We need to consider all the various races and types of humans. What if on Mars, or another planet, some other animal, other than primates, developed a larger cranium and brain...for example, giant insects or intelligent reptilians, like in the movies. We need to allow ourselves a wide latitude on our expectations, or we may look at them and not SEE them.

The most recent photos by Curiosity Rover on Mars seem to show some trace evidence that humans exist on Mars. (It's important for readers to understand that "evidence" is not "proof"). A humanlike footprint, for example, is "evidence". As strange as this may seem, (and I say "strange", considering

the generally presumed concept of a "lifeless" Mars), there are photos that suggest a dam, pipes, a church, and roads. These seemingly manmade features could likewise be natural formations or photographic distortions.

In evaluating the question "are there humans on Mars" one should look at all the evidence, prior to making an informed decision or conclusion. If there are photos that show those types of things, then I believe that is one form of evidence of human existence, or possibly <u>prior</u> existence. A determination needs to be made of whether the photos show actual humans, or do they just show something that appears to be humans. I will try to explain and show how these things occur in the photos, and then present my opinion.

THE DAM

The feature indicated in the following photo looks like a dam. It could just be a bridge, or, that is, a continuation of a road which passes across a valley, (which, without culverts, becomes a dam if there is any water behind it). It seems to align with the horizontal geologic layering of the hills around it. It may be level or it may be slightly sloped, that is not clear. But it is a perfectly straight line for hundreds of meters. The "hill" in the center of the photo is estimated by NASA to be three hundred meters wide. Using the hill as a scale makes the dam over a half a kilometer long.

ROADS

A road appears to be a ramp approaching the dam, constructed, rather than naturally-formed. The road appears to have a constant slope, from the lowest level of the valley, up to the dam. The portion of the road visible in the photo is approximately three kilometers long. This photo gives a sense of the distances involved in the Curiosity photo. The rover is about ten kilometers from the dam. The road passes over a waterway with a fairly long bridge, demonstrating engineering ability.

In the photo below, there appears to be a road or two that follow the edge of the water. These could be paths worn by animals or they could be constructed roads. It could be an indication of a higher water level in the recent or ancient past. This photo shows several types of terrain. There is a rocky area in the foreground. There is a dried grass area beyond that. There is a river or pond, with banks. And then there is a dry-grass-covered hill in the background.

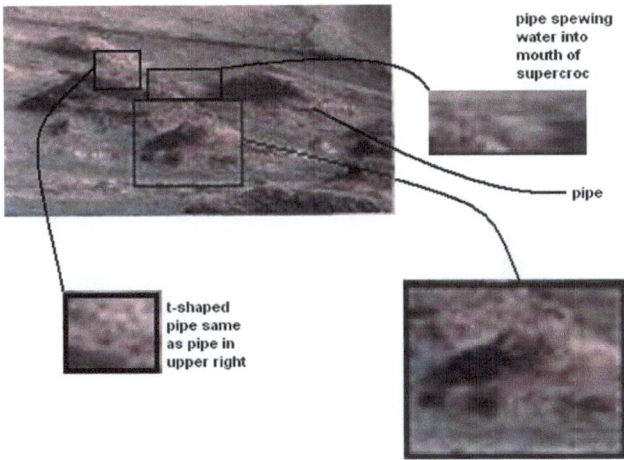

pipe spewing water into mouth of supercroc

pipe

t-shaped pipe same as pipe in upper right

PIPES

This photo shows the area near the bottom of the dam. There appears to be a pipe shaped like a "T" which comes out of the ground vertically, then is spewing water horizontally a great distance. A super croc or a similar animal seems to be open-mouthed, drinking the water. The animal is about fifty meters long, judging from the 300 meter "hill" farther to the left and out of the picture. (The hill, by the way, is probably not a hill.) There is another similar T-shaped pipe coming out of the ground nearby and beyond. A horizontal pipe lies near, and in front of, the tail of the animal which is drinking the water. Additionally, there is

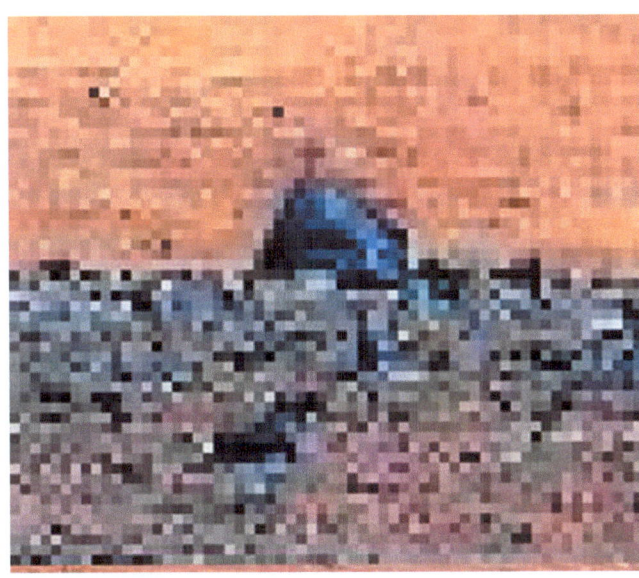

CHURCH

This dark object looks like a building with a cross on top. It might not be what it appears to be. The cross is only visible on this version of the photo...many other versions of the same photo do not have a cross. But this "object" resembles, in shape, many of the "rocks" in photos. The rocks with a shape like this could be rear-ends of animals...dinosaur-like animals such as stegosaurs. This one is the size of a house...and the shape of a house.

a small (ten meters long) crocodile-like animal posed in defense mode with mouth open, as a huge snake appears to be in the act of striking it.

TANK

At the base of the dam, there appears to be something metallic and cylindrical, like a large pipe or tank. By comparing it to the known dimension of the 300-meter wide "pointy mound" next to it, it can be estimated to have a diameter of about ten meters, and length of fifty meters.

pipe or tank

HUMAN IMAGES

Human shapes are detectable in some of the Curiosity photos. There is a problem with all of the human images in that they all seem quite vague and blurry, sometimes transparent or ghostlike in nature. Partly this is due to the vast distances from the cameras, (10 km), and the plastic dust cap. Another problem is that one photo seems to have many versions, probably because there are multiple cameras involved taking multiple pictures. So each version of each photo is different, and no two mosaics seem to match when zoomed in completely. It's essentially like taking many portions of pictures over time, in which the subjects are moving around...multiple exposures. But strangely, this only occurs with humans. It's quite mystifying.

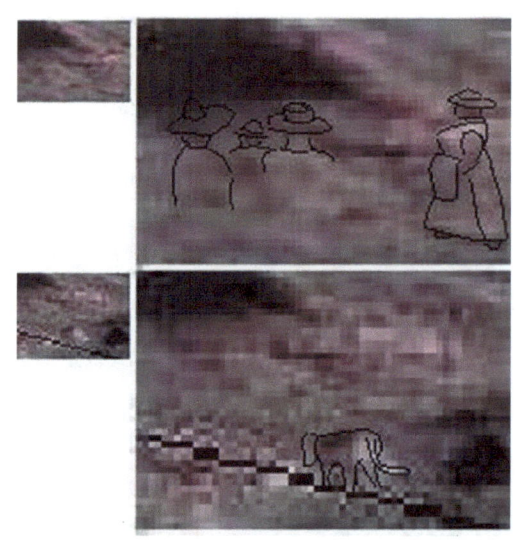

40

Nevertheless, it is sufficiently clear to make out the forms of some of the people and to ascertain their basic nature. They seem to be clothed in much like western attire. Women wear full length dresses, while sometimes-shirtless men wear pants. Both men and women wear hats like sombreros. The people seem to be dark-skinned, and heavy or large in size, well-fed. The above photos, which represent two versions of the same photo, seem to show a large number of people at some kind of community or family celebration. They don't seem to be heavily dressed. Is it summertime on Mars? There is what looks like a dog like a German Sheppard on the ledge in the foreground. Colors of the people are faded or dull, like olive drab, although the large woman in the upper version seems to have a red and white print dress, carrying a wrap on her arm. These features are not what I or anyone would expect on Mars. Rather, it appears incredibly Earthlike and modern. This could be in Africa or Polynesia on Earth. Scaling the people, by using the known 300 meter "hill" in the same photo, these people are twenty meters tall. This really depends on the accuracy of the scale guidance from NASA. Could they be off by a factor of ten? Or are people really ten times the size of humans on Earth? I cannot answer that question at this time.

HUMAN AT GLENELG

There is a strikingly human form in the satellite photo taken near Glenelg. It looks something like a man on top of a centipede, as if climbing over it. Such monotone images could be easily mistaken. In fact one needs to wonder if it is really what it looks like. Why would a human be so large? The scale of this photo is known, and according to the scale shown on the photo, this person would be twenty meters tall. And why would he climb over a centipede? Or is it all just rock formations?

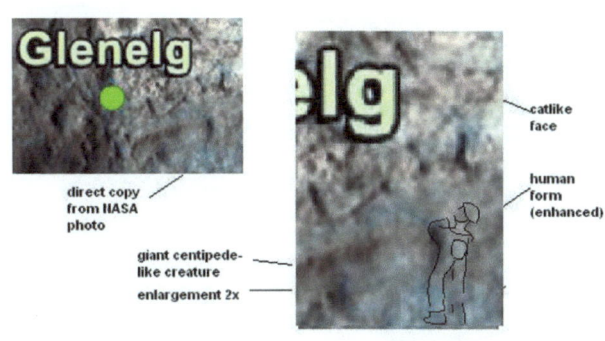

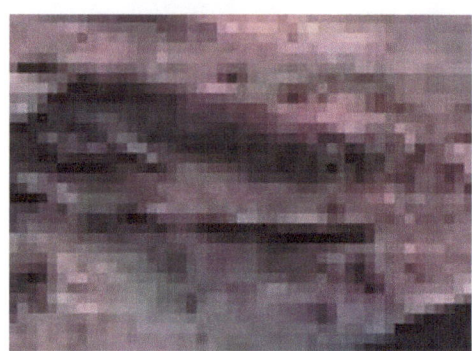

The photo above shows several people standing on a log looking into a crocodile pit. Based on NASA's estimated 300 meter-wide hill in the same photo, the tallest figure in the picture scales 20 meters tall. The sketch at right shows a comparison with Earth humans, 2 meters tall.

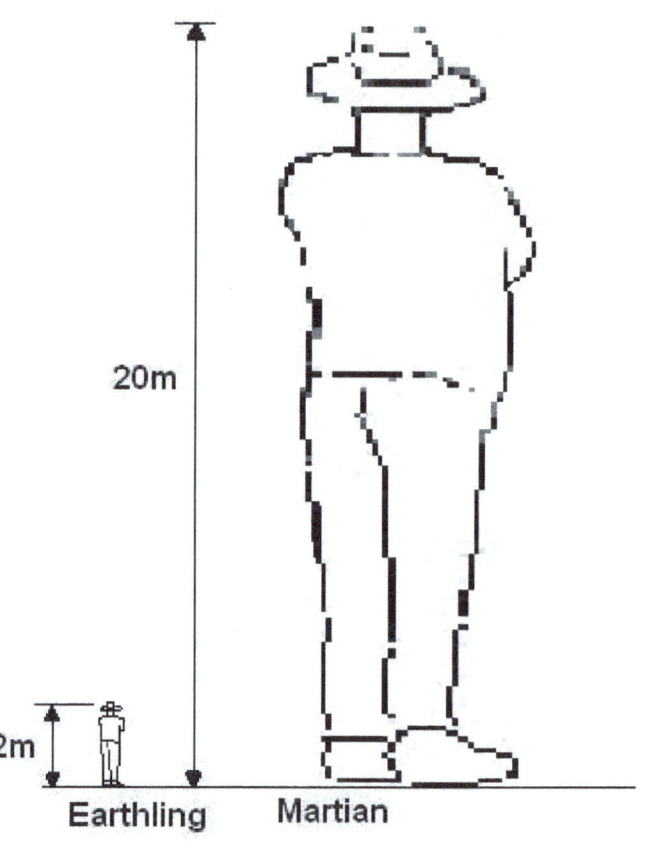

20m

2m

Earthling   Martian

ARE THERE HUMANS ON MARS?

So, having reviewed the evidence, can any conclusions be drawn? There is the dam, the road, the pipes, the culvert, and human images. Are there humans on Mars, or not? There is difficulty in answering that question, even after reviewing the above photos. Each photo is vague, blurry, questionable, and unclear. It really is up to the viewer as an individual to form a conclusion based on this scanty evidence. But the sum of many things is greater than one individually. My personal opinion, having studied the question in depth, and for several days, weeks, and months now, is that humans do exist on Mars, much like modern humans; however, they are gigantic in comparison to Earth humans. I would say, however, that I believe the Curiosity rover will cross paths with waterways, animals, plants, and humans, in the very near future, and the question will be answered conclusively. The sand is rapidly emptying to the bottom of the hourglass. We need to prepare for interactions with humans on Mars.

In the Curiosity photo taken October 15, 2012, a human head can be seen in the sand at the site NASA calls Rocknest. This appears to be a man with a diving suit, with gloves. I suspect he is part of a search team, with canoes and dogs, possibly looking for someone lost in the dry quicksand.

In addition to the several photos of humans in this part, see also the photo of the Supergoose head in the ground on page 30. A human form can be seen standing beside the giant head. This photo is the clearest photo of a human, and even the person's straps for a backpack can be seen, in addition to his skin color and hair color.

PART 6: CONCLUSIONS

At the time of this writing, Mars is considered lifeless, liquid waterless, and desolate, in the eyes of science. The images presented herein could change that concept completely.

The photos from Mars show water and life including plants, birds, and many animals, possibly humans. Some are exactly like those on Earth. Some are even pre-extinction dinosaurs. Some are gigantic, similar to those on earth but many times larger. Still other animals are totally different from those on Earth. But there is no longer a question about whether there is life on Mars. There IS life on Mars...in abundance.

The photos from NASA's rovers seem to show many things that could be regarded as evidence of humans on Mars. Roads, a dam, pipes, a footprint, a church with a cross, as well as ghostlike human images. But, with each of these photos having a questionable nature to it, I won't go so far as to say that this "proves" anything; however, I will say that the possibility of human life on Mars is now very likely, in my opinion. Earth humans need to prepare to meet their neighbors on Mars, as a meeting is inevitable. We need to learn their language. We need to take equipment that would enable communication, whether visual and/or audible. Simulation of the atmosphere on Earth and testing of voice communications gear in the thinner air is needed.

We can forget about taking enough oxygen to last for not only the nine month voyage, but for months on Mars. We need to think about acclimation for Mars exploration by humans. Acclimate

before leaving, by living in simulation chambers, gradually lowering the oxygen levels, or possiblly, during the voyage. The atmosphere needs to be duplicated on Earth in a test chamber, and then tested for habitability by animals, then humans. Forget about taking food for the duration on Mars. Food may be readily available through communication with Martians, if that becomes possible. Think fish or goose eggs.

Humans did not exist until over four million years ago, or at least that is what anthropology tells us. Therefore, there might have been an Earth/Mars exchange within the past four million years. In fact, if the humans on Mars are similar to and evolved from humans on Earth, modern humans, then the exchange had to have been after the advent of the modern human, which could be within the past sixty thousand years. Consider:

> Sand dollars did not exist over fifty million years ago, therefore, there must have been an Earth/Mars exchange within the past fifty million years. It might have occurred at the same time, sixty thousand years ago, or it might have occurred since 50 million years ago...or both.

> Velociraptors lived on Earth 75 to 71 million years ago, and therefore there must have been an Earth/Mars exchange at approximately that same period.

> Supercrocs became extinct and therefore did not exist 112 million years ago, and therefore, there must have been an Earth/Mars exchange prior to that time...say roughly around 120 million years ago.

> Stegasaurus lived 155 to 150 million years ago, therefore, there must have been an Earth/Mars exchange prior to that time...perhaps 160 million years ago.

My thinking of Earth/Mars exchanges come, in part, from Emanuel Velikovsky's theories in his book, "Worlds in Collision", (which I realize are generally disregarded by modern scientists, but then, there is significant historical information there). In that book, he explains how ancient writings tell about times that the planet Mars approached Earth. Velikosky theorizes that comets disturbed the orbits of Mars and Venus, causing them to temporarily have more elliptical shape, which then interfered with Earth, resulting in great chaos and historical events like earthquakes, floods, meteor showers, and change of calendar (e.g. 787 BC). Another possible repercussion of this idea is that perhaps the nephelim, the giants of the Bible, were actually people from Mars who happened to fall to Earth during a Mars Earth-exchange. I'm presenting these ideas for further exploration.

I'm suggesting that it happened several times, perhaps on a regular basis, (such as due to a comet or Sun-partner), every forty million years or so, causing interchange of life forms from Earth to Mars. Perhaps that would explain the huge canyon-like rift along Mars' equator. Perhaps it is not a crack but a scratch from Earth contact. If so, there should be some Martian soil and/or life forms on Earth, most likely near the equator, as well as Earth life forms on Mars near the rift. Velikovsky says that a mountain moved near Jerusalem in 787 BC.

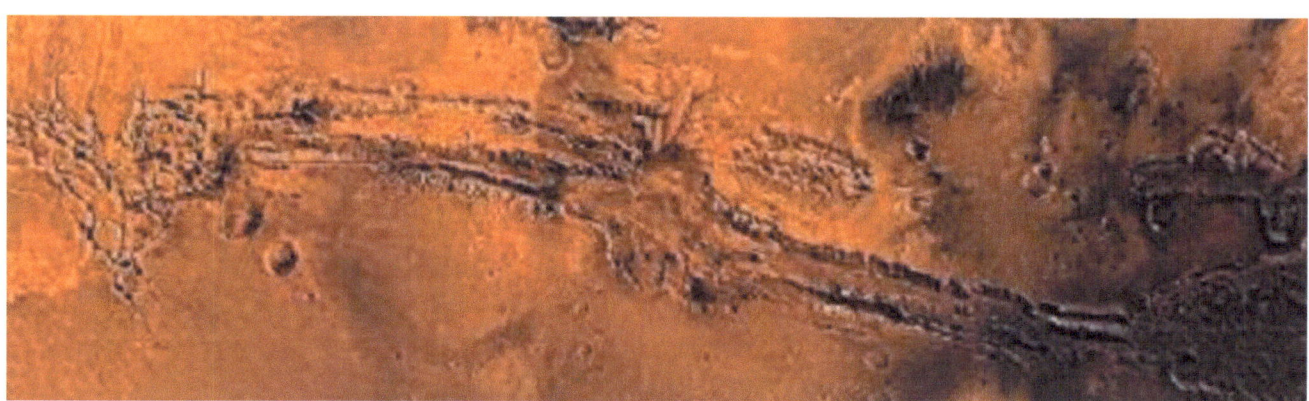

www.ingramcontent.com/pod-product-compliance
Lightning Source LLC
Chambersburg PA
CBHW051100180526

45172CB00002B/711

* 9 7 8 1 4 7 9 7 3 3 7 8 1 *